D1527412

ATMOSPHERIC EFFECTS

OF AVIATION

A Review of NASA's Subsonic Assessment Project

Panel on Atmospheric Effects of Aviation
Board on Atmospheric Sciences and Climate
Commission on Geosciences, Environment, and Resources
National Research Council

NATIONAL ACADEMY PRESS
Washington, D.C. 1999

NOTICE: The project that is the subject of this report was approved by the Governing Board of the National Research Council, whose members are drawn from the councils of the National Academy of Sciences, the National Academy of Engineering, and the Institute of Medicine. The members of the committee responsible for the report were chosen for their special competences and with regard for appropriate balance.

Support for this project was provided by the National Aeronautics and Space Administration under grant NASW-4938 order No.109. Any opinions, findings, and conclusions or recommendations expressed in this publication are those of the author(s) and do not necessarily reflect the views of the above-mentioned agency.

Additional copies of this report are available from:
National Academy Press
2101 Constitution Ave., NW
Box 285
Washington, DC 20055
800-624-6242
202-334-3313 (in the Washington, D.C., metropolitan area)
www.nap.edu

PANEL ON ATMOSPHERIC EFFECTS OF AVIATION THAT PREPARED THIS REPORT

ALBERT J. KAEHN (*Chair*), Brigadier General, U.S. Air Force, retired
*JACK G. CALVERT, National Center for Atmospheric Research, Boulder, Colorado
GEORGE F. CARRIER, Harvard University (emeritus), Boston, Massachusetts
*ANTONY D. CLARKE, University of Hawaii, Honolulu
DIETER H. EHHALT, Institut für Atmosphärische Chemie, Jülich, Germany
*DAVID J. ERICKSON III, National Center for Atmospheric Research, Boulder, Colorado
*CLAIRE GRANIER, Université Paris, France; National Oceanic and Atmospheric Administration and Cooperative Institute for Research in Environmental Sciences, Boulder, Colorado
EDWARD M. GREITZER, Massachusetts Institute of Technology, Cambridge
*PHILIPPE MIRABEL, Université Louis Pasteur, Strasbourg, France
MICHAEL OPPENHEIMER, Environmental Defense Fund, New York, New York
W. GEORGE N. SLINN, Cascade Scientific Research Corporation, Richland, Washington
*KNUT H. STAMNES, University of Alaska, Fairbanks

Staff

ELLEN F. RICE, Study Director (through 9/98)
LAURIE S. GELLER, Program Officer
TENECIA A. BROWN, Senior Program Assistant

---

*Members of the subsonic working group

iii

The National Academy of Sciences is a private, nonprofit, self-perpetuating society of distinguished scholars engaged in scientific and engineering research, dedicated to the furtherance of science and technology and to their use for the general welfare. Upon the authority of the charter granted to it by the Congress in 1863, the Academy has a mandate that requires it to advise the federal government on scientific and technical matters. Dr. Bruce M. Alberts is president of the National Academy of Sciences.

The National Academy of Engineering was established in 1964, under the charter of the National Academy of Sciences, as a parallel organization of outstanding engineers. It is autonomous in its administration and in the selection of its members, sharing with the National Academy of Sciences the responsibility for advising the federal government. The National Academy of Engineering also sponsors engineering programs aimed at meeting national needs, encourages education and research, and recognizes the superior achievements of engineers. Dr. William A. Wulf is president of the National Academy of Engineering.

The Institute of Medicine was established in 1970 by the National Academy of Sciences to secure the services of eminent members of appropriate professions in the examination of policy matters pertaining to the health of the public. The Institute acts under the responsibility given to the National Academy of Sciences by its congressional charter to be an adviser to the federal government and, upon its own initiative, to identify issues of medical care, research, and education. Dr. Kenneth Shine is president of the Institute of Medicine.

The National Research Council was organized by the National Academy of Sciences in 1916 to associate the broad community of science and technology with the Academy's purposes of furthering knowledge and advising the federal government. Functioning in accordance with general policies determined by the Academy, the Council has become the principal operating agency of both the National Academy of Sciences and the National Academy of Engineering in providing services to the government, the public, and the scientific and engineering communities. The Council is administered jointly by both Academies and the Institute of Medicine. Dr. Bruce M. Alberts and Dr. William A. Wulf are chairman and vice-chairman, respectively, of the National Research Council.

Preface

This report is the fourth assessment provided to the NASA Atmospheric Effects of Aviation Project (AEAP) by the National Research Council's Panel on the Atmospheric Effects of Aviation (PAEAN). The AEAP has two parts, the Atmospheric Effects of Stratospheric Aircraft (AESA) component and the Subsonic Assessment (SASS) component. The SASS project, which is the subject of this report, has the goal of assessing the environmental impacts of both the current and future fleets of subsonic civil transport aircraft. Begun in late 1993, SASS has sponsored field campaigns, laboratory studies, and atmospheric modeling efforts. The panel focuses here on SASS's recent self-evaluation, *Atmospheric Effects of Subsonic Aircraft: Interim Assessment Report of the Advanced Subsonic Technology Program*, and on the project's strategic plan for the next few years.

Panel members were selected to provide expertise in field observations, laboratory chemistry, atmospheric dynamics and modeling, aircraft engines, and climate. NASA's charge to the panel is to evaluate the appropriateness of the SASS research plan, to appraise the project-sponsored results relative to the current state of scientific knowledge, to identify key scientific uncertainties, and to suggest research activities likely to reduce those uncertainties. PAEAN has already published a series of three interim evaluations of AEAP: *An Interim Review of the Subsonic Assessment Project* and *An Interim Assessment of AEAP's Emissions Characterization and Near-Field Interactions Elements* in 1997, and *An Interim Review of the AESA Project: Science and Progress* in 1998. A final evaluation of AESA is currently under way. The primary audience for these reports is the program managers and scientists affiliated with AEAP, although in some cases the topics discussed will likely be of interest to a wider audience.

The panel appreciates the dedication of its longtime study director Ellen Rice, as well as the support of its new staff officer Laurie Geller. We are also grateful to the researchers and managers who have provided briefings and reports to keep the panel apprised of the progress of SASS and related research programs.

This report has been reviewed by individuals chosen for their diverse perspectives and technical expertise, in accordance with procedures approved by the NRC Report Review Committee. The purpose of this independent review is to provide candid and critical comments that will assist the authors and the NRC in making the published report as sound as possible and to ensure that the report meets institutional standards for objectivity, evidence, and responsiveness to the study charge. The content of the review comments and draft manuscript remain confidential to protect the integrity of the deliberative process. We thank the following individuals for their participation in the review of this report:

Alan Epstein, Massachusetts Institute of Technology
Thomas Graedel, Yale University
Matthew Hitchman, University of Wisconsin
Sonia Kreidenweis, Colorado State University
Shaw Liu, Georgia Institute of Technology
John Seinfeld, California Institute of Technology

While the individuals listed above have provided many constructive comments and suggestions, responsibility for the final content of this report rests solely with the authors and the NRC.

Albert J. Kaehn, Jr.
Former PAEAN Chair

Contents

Executive Summary

Aviation is an integral part of the global transportation network, and the number of flights worldwide is expected to grow rapidly in the coming decades. Yet, the effects that subsonic aircraft emissions may be having upon atmospheric composition and climate are not fully understood. To study such issues, NASA sponsors the Atmospheric Effects of Aviation Program (AEAP). The NRC Panel on Atmospheric Effects of Aviation is charged to evaluate AEAP, and in this report, the panel is focusing on the subsonic assessment (SASS) component of the program. This evaluation of SASS/AEAP was based on the report *Atmospheric Effects of Subsonic Aircraft: Interim Assessment Report of the Advanced Subsonic Technology Program* (Friedl, 1997), on a strategic plan developed by SASS managers, and on other relevant documents.

While SASS has made significant progress over the last couple of years, significant uncertainties remain, particularly with regard to the chemical and radiative impacts of particles, including sulfur and carbon aerosols, contrails, and modification of natural cirrus clouds. In this report, the panel reviews recent scientific developments that it sees as the most relevant to AEAP and notes many specific areas where more research attention would be particularly useful. Overall the panel identified the following as its highest priority concerns and recommendations:

- **Impacts of heterogeneous chemistry on ozone**. Heterogeneous reactions couple strongly with gas-phase processes and may significantly affect ozone concentrations; thus, assessments of the impacts of aviation on ozone levels must

1

include the uncertainties associated with the climatology of aerosols and the heterogeneous processes occurring on these aerosols.

• **Emissions characterization**. More aircraft engine emissions characterization studies, particularly for particles, sulfur oxidation products, and hydrocarbons would be worthwhile. To accurately interpret field measurements of these exhaust species, methods need to be developed to distinguish in situ between emissions from aircraft and those from other sources.

• **Aerosol climatology.** Aircraft particulate emissions must be assessed in the context of the ambient particle field. It is recommended that SASS assemble currently available baseline data on the ambient particle fields in the upper troposphere and lower stratosphere, so that the relative effects of aircraft-emitted particles can be properly gauged.

• **Troposphere-stratosphere exchange.** Cross-tropopause transport processes are not well enough known to permit appropriate representation in models, and thus need to be an important focus of AEAP analyses. It is important that SASS and AESA sort out this composite assessment with whatever type of collaboration will be most effective.

• **The Global Modeling Initiative**. Carefully coordinated sensitivity studies are needed to test the dependence of the Global Modeling Initiative (GMI) results upon the parameterizations of sub-grid processes, stratosphere–troposphere exchanges, and the chemical schemes selected. Likewise, the tropospheric components of the GMI should participate in international model-intercomparison studies. SASS's large commitment to the GMI can be justified only if heterogeneous processes (including the formation and chemistry of aerosols, contrails, and cirrus) can be incorporated into the model framework in the near future; otherwise, it may be wise to invest in alternative modeling tools. Finally, the GMI efforts should be more closely coordinated with the SASS-sponsored climate modeling studies.

• **Planning of field campaigns**. The results of the recent field campaigns SUCCESS and SONEX should be used to guide the planning and conduct of SASS's next campaign, helping to pinpoint the most important scientific questions and the most effective measurement methodologies.

• **Cooperation with other research programs.** SASS experimental research should focus as much as possible directly on the effects of aircraft emissions; the importance of these perturbations, however, can only be evaluated in the context of the natural distribution and variability of the relevant chemical species, as well as the magnitude of other anthropogenic perturbations. SASS should give much higher priority then, to exploring how data accumulated by other programs can be used to provide some context in which aircraft effects can be evaluated. We recommend that SASS devote more of its time and money to establishing cooperative efforts with other research programs, both within and beyond NASA, and to evaluating the possible applicability of datasets and information gathered by others. Likewise, SASS should be preparing to use the existing and anticipated data from satellite platforms, identifying the specific data products that will be relevant to evaluating the impacts of aircraft emissions.

1

Context For This Review

INTRODUCTION

Aviation is an integral part of the global transportation network. The number of commercial aircraft flights has grown steadily for decades and is expected to increase even more rapidly in the years to come, particularly in developing countries. The impacts of this activity on atmospheric composition and climate are not yet entirely clear. Researchers are only now beginning to quantify the likely effects of aircraft exhaust, which includes carbon dioxide (CO_2), water (H_2O), oxides of nitrogen (NO_x), carbon monoxide (CO), hydrocarbons (HCs), sulfur (S), and soot. Assessing the atmospheric effects of aviation requires understanding processes with scales that range over orders of magnitude—from microphysical details of particle production in the engine to global climatic changes.

Over the last decade, most attention has been focused on the questions of how aircraft emissions (principally NO_x) could interact with species present in the atmosphere to alter the concentration and distribution of ozone, and how these changes, as well as the addition of CO_2 and water vapor, could influence climate. The focus of concern has recently shifted to other impacts that are less direct but may ultimately be more important. For instance, contrails, which are formed by water vapor and particulate matter in aircraft exhaust, may increase the amount of cirrus clouds in the atmosphere, thereby influencing radiative forcing. Likewise, ozone concentrations could be affected through the heterogeneous chemical reactions that take place on the surface of the particles emitted or formed in aircraft wakes and on naturally occurring clouds.

All these concerns take on added importance because of the ongoing international climate change treaty negotiations. The Kyoto Protocol to the United Nation's Framework Convention on Climate Change specifically mentions the need to limit emissions of greenhouse gases from aviation. A determination that aviation is a significant contributor to anthropogenic effects on climate could have far-reaching regulatory and economic implications for the airline industry.

The NRC Panel on Atmospheric Effects of Aviation (PAEAN) is charged with evaluating the direction and progress of the Subsonic Assessment (SASS) component of NASA's Atmospheric Effects of Aviation Project (AEAP). PAEAN was asked to evaluate the appropriateness of the research plan, to appraise the project-sponsored results relative to the current state of scientific knowledge, to identify key scientific uncertainties, and to suggest research activities likely to reduce those uncertainties. In doing so, the panel has focused principally on SASS's most recent self-assessment, *Atmospheric Effects of Subsonic Aircraft: Interim Assessment Report of the Advanced Subsonic Technology Program* (Friedl, 1997), and on the SASS strategic plan that governs the next few years' research. The complementary progress described in the *European Scientific Assessment of the Atmospheric Effects of Aircraft Emissions* (Brasseur et al., 1997), as well as other appropriate literature, were also taken into consideration.

The panel had the benefit of briefings by two of SASS's project scientists at its November 1997 meeting, followed by discussions with them about SASS's objectives and factors that affect the project's ability to achieve those objectives. The panel then decided that the appropriateness of the SASS research plan needed to be examined in the context of all related NASA research activities; thus, at the panel's January 1998 meeting, representatives of programs in NASA's Office of Earth Science (formerly Mission to Planet Earth) briefed the panel on other ongoing and planned research activities. These discussions of NASA's Tropospheric Chemistry Program, Upper Atmosphere Research Program, Radiation Science Program and atmospheric chemical modeling activities were extremely useful to the panel's deliberations, and influenced the recommendations given in this report.

THE SASS STRATEGIC PLAN

To implement a project like SASS, it is essential to have an operational framework for characterizing objectives, establishing goals, and prioritizing activities. In its earlier review of SASS (NRC, 1997a), the panel requested that AEAP staff provide an explicit plan that would establish a context for reviewing and interpreting the project's activities. A strategic plan was drawn up and presented to the panel at its November 1997 meeting. It has since been modified somewhat. This plan provides both a vision and an approach to which the panel can compare SASS's specific efforts. The key elements of SASS's strategic plan are:

• **Goal:** Develop a scientific basis for assessment of the atmospheric impact of subsonic aviation, particularly commercial aircraft cruise emissions.

- **Objective:** Provide a measure of the change in radiative forcing due to changes in ozone (O_3), carbon dioxide (CO_2), water vapor (H_2O), and aerosols in the upper troposphere and lower stratosphere by emissions from both present and future subsonic aircraft.

- **Minimum Success Requirement:** a quantitative measurement of the changes in O_3, CO_2, H_2O, and aerosols resulting from subsonic aircraft emissions, with quantified uncertainties.

- **Key Scientific Issues:**
  1. What are current and future emissions from aircraft?
  2. What chemical and physical processes in the atmosphere could be perturbed by aircraft emissions?
  3. Are atmospheric observations consistent with the current understanding of aircraft emissions-related chemistry and physics?
  4. What are the predicted ozone changes and climatic impact associated with aviation?
  5. What are the uncertainties in these predictions?

- **Major Science Objectives for 1999 (in priority order):**
  1. Reduce uncertainty in evaluation of present-day tropospheric O_3 change due to aircraft NO_x.
  2. Constrain estimates of indirect radiative impacts of soot and sulfur emissions.
  3. Quantify radiative forcing from contrails.
  4. Assess impact of subsonic aircraft emissions on stratospheric O_3.
  5. Calculate changes in CO_2 and O_3 radiative forcing from future aircraft fleet (with IPCC).
  6. Provide robust limit on potential impacts of CO and non-methane hydrocarbon emissions on O_3 chemistry.

- **Major Science Objectives for 2001 (in priority order):**
  1. Begin to quantify indirect radiative impact of soot and sulfur emissions.
  2. Calculate O_3 changes and climate forcing from aircraft NO_x for the current and future fleets, with a factor of two as the goal for uncertainty level.

SASS's overall approach is designed to lead to formal assessment reports that can be used to guide decisions related to the development of a new generation of subsonic aircraft. The strategy taken by AEAP in general is to link the information gained from emissions characterization and laboratory studies to interpretation of near-field interactions and aircraft operational scenarios. This information is intended to be used as input to models, and as a means of interpreting atmospheric observations. The results will constitute a basis for the development of the assessment reports.

PAEAN's assessment of SASS was also made in light of an advisory it received from Howard Wesoky, then with NASA's Office of Aeronautics and Space Transportation Technology (OASTT). He made the following points about his perception of the appropriate scope and responsibility of AEAP, particularly with regard to the subsonic assessment component: the impacts of subsonic aviation do not appear to be unique and are likely to be small relative to total anthropogenic influences on the atmosphere; obtaining a more accurate assessment of aviation's impacts will require many more years of research and considerably greater investment than is possible from the Advanced Subsonic Technology and OASTT programs; a prudent management approach to this assessment area may be to recognize that within NASA the basic research necessary to advance understanding is the responsibility of the Office of Earth Science.

These issues clearly have a substantive bearing on the SASS program, and thus on the deliberations of the panel. The confluence of the strategic plan and the statements listed above can be perceived as addressing a key point that PAEAN made in its *Interim Review of the Subsonic Assessment Project* (NRC, 1997a): "It is essential that AEAP apply its research funds where they are most likely to reduce the major uncertainties."

The Goals and Organization of This Review

PAEAN's evaluations, and the recommendations given in this report, are aimed primarily at helping SASS make the greatest possible progress before its final assessment report, slated for the year 2001. The panel raises issues that it feels deserve more attention than they are receiving at present, and it suggests where research performed outside SASS may help bound uncertainties. However, it will very likely take until well beyond 2001 to identify with reasonable certainty all the processes that may be important in quantifying the effects of the present or future subsonic fleets on the atmosphere. Some longer-term goals and strategies are therefore discussed as well.

The panel feels that, in general, SASS has correctly identified the most important scientific issues, and the project's major objectives are all of considerable scientific interest. The SASS strategic plan represents a significant step forward in clarifying the goals and strategies of the program. The panel has some concerns, however, that the current emphases may not be properly balanced, and that the program's goals may not be most efficiently and rapidly achieved with the present approach.

Chapter 2 contains a discussion of the different scientific issues that form the basis of SASS's investigations, and Chapter 3 continues this discussion, focusing on the issues directly related to modeling. It should be noted that these discussions are not meant as a comprehensive review of all issues involved with subsonic aviation's atmospheric impacts[1] ; rather, they are limited to the particular issues that the panel feels present the greatest remaining sources of uncertainty and may present the most important opportunities for SASS to make progress in reducing uncertainties. Chapter 4 contains a list of the panel's highest priority concerns and recommendations, covering both scientific and management related issues.

---

[1]A comprehensive assessement of the atmospheric impacts of aviation is being prepared by the Intergovernmental Panel on Climate Change (IPCC). Since this IPCC report is not yet final, the PAEAN report neither draws upon or refers to the IPCC effort.

2

Discussion of the Science Issues

The topics covered in this chapter are grouped into four general areas: gas phase processes, particle-related processes, atmospheric transport, and climate impacts. For each of these general areas, the panel looks at the findings of both the NASA and European assessment reports and raises issues judged to merit most attention in light of SASS's objectives and schedule. In some cases, the issues are the same as those highlighted by SASS's strategic plan; in other cases they suggest a shift in priorities. It should be noted that research on many of these issues is progressing rapidly, and the discussions presented here can really only provide a concise snapshot of the state of the science when the panel's analyses were written.

GAS-PHASE EMISSIONS, CHARACTERIZATION, AND CHEMISTRY

A central focus of SASS investigations has been to quantify the effects of aircraft emissions, particularly NO_x, on atmospheric ozone levels. (The panel's interim review of SASS contains a description of the basic processes involved in ozone photochemistry and will not be reviewed here.) Some highly uncertain issues remain, such as the lightning NO_x source, and the need for more upper troposphere/lower stratosphere trace gas measurements to constrain model estimates. In recent years, though, some significant progress has been made. For instance, trace gas measurements made during field campaigns such as SUCCESS, SONEX, and STRAT[2] have led to valuable new insights into the role

---

[2]SUCCESS (Subsonic Aircraft: Contrail and Cloud Effects Special Study), SONEX (SASS Ozone and Nitrogen Oxide Experiment), STRAT (Stratospheric Tracers of Atmospheric Transport).

of NO_x, HO_x, and HCs in determining the photochemistry of ozone in the upper troposphere. Also, detailed NO_x source inventories, laboratory investigations of key reactions, and emission characterization studies have helped provide input data for assessment models.

The chemistry as outlined in the Brasseur et al. (1997) report is shown in somewhat more detail than that given in the Friedl (1997) report, but where comparisons can be made it appears that there is general agreement on the specific chemistry employed by both groups. It is likely that the chemical mechanisms chosen by the NASA and European groups are very similar, since both U.S. and European groups rely heavily on the two existing major kinetic data reviews, which are very similar in their recommendations (DeMore et al., 1997; Atkinson et al., 1997). In general, SASS's research objectives related to the homogeneous chemistry used in the project's modeling efforts do not appear to be missing any critical elements. However, there are some issues the panel feels may deserve more attention, as discussed in the following sections.

Aircraft Emission Issues

The nature and extent of the chemistry included in the assessment models should ideally be based on our knowledge of the aircraft emissions under actual flight conditions. In the absence of such knowledge, choices are made on the basis of measurements made during bench tests of aircraft engines. Observations of exhaust plume components seem to have been restricted to those components that are believed, a priori, to be important (e.g., NO_x, CO, SO_x, soot, aerosols). Friedl (1997) suggests strongly that this information is still very incomplete. As discussed below, there are some issues related to the emissions of hydrocarbons and nitrogen-containing compounds that may deserve particular attention; discussion of sulfur-compound emissions is reserved for the 'aerosols' section of this report.

Although current jet-engine bench tests indicate that aircraft engine combustion is very efficient under many operating conditions, combustion in any engine is never "complete." Some unburned fuel is found in jet-aircraft exhaust, particularly when a fuel-rich mixture is used in the engine. Friedl (1997) notes on p. 36, "Aircraft exhaust is known to contain a large number of C_2-C_{17} species, although the relative amounts are not well established." This means that assessments must consider the effects on ozone chemistry of unburned hydrocarbons and hydrocarbon oxidation products in the exhaust plume.

Our knowledge of the hydrocarbons present in current jet fuels is reasonably good, but some important minor components remain ill-defined. JP-8 fuel, which is used in many units of the U.S. Air Force, complies with a set of specifications that are essentially identical with those of civilian aviation fuel (JA-1), except for fuel additives required by the JP-8 specification (CRC, 1984). Mayfield (1996) reports that on average the JP-8 jet fuel mixture consists of about 80.4% alkanes (mostly C_{10} to C_{14}), 17.9% aromatic hydrocarbons, and 0.8% alkenes. The normal (straight-chain) alkanes, n-C_9H_{20} to n-$C_{15}H_{32}$ make up 34% of the carbon in the fuel. The properties given by Schumann et al. (1996) for European jet fuel reflect a similar composition: largely C_{10} to C_{14} alkanes and C_8 to C_{10} alkyl-substituted benzenes and napthlenes.

The atmospheric lifetime of hydrocarbons in this mass range is determined largely by reactions with OH radicals; it is relatively short for the conditions typical of the upper troposphere and lower stratosphere, but some of the initial products of oxidation of the large alkanes (e.g., the ketones) have longer lifetimes than the original alkanes. The extent of oxidation of the fuel is, among other factors, dependent on the air-to-fuel ratio used by the aircraft. To our knowledge, no analyses of trace hydrocarbon concentrations in the C_{10}-C_{14} range or of their oxidation products in the air-traffic corridor have been reported. More tests for these NMHC emissions should be made in the flight corridor regions. Not only may these species serve as additional tracers for aircraft-exhaust dispersion, but they may also represent an important input into model simulations of ozone chemistry in the troposphere.

These questions may be particularly relevant with respect to military aircraft, as Friedl (1997) notes that military aircraft have been calculated to account for a disproportionately large fraction (>30%) of HC and CO aircraft emissions. As suggested, further work to lower the present uncertainties in the military database is needed. This will require detailed consideration of the issues raised above and further modeling efforts devoted to the problem.

An important factor in designing the atmospheric transport/chemistry models used in the NASA and European assessments (Friedl, 1997; Brasseur et al., 1997) is the degree of chemical complexity that must be included to attain a credible accuracy in atmospheric simulations. The need to use a realistic composition of hydrocarbon reactants in fuel is borne out by model-sensitivity tests, as Friedl (1997) reports on p. 82: "Removing hydrocarbon chemistry from the model reduces the sensitivity of ozone to aircraft NO_x emissions appreciably (from 0.97% to 0.66% globally and annually averaged)." In view of these tests, it is surprising that Friedl concludes on p. 36, "Although the impact of NMHCs from subsonic aircraft emissions is likely to be small, no serious effort to accurately simulate these effects has been undertaken to date." More model-sensitivity studies are required to justify any conclusions about the effects on the ozone column of hydrocarbon reactions in aircraft exhaust.

Similarly, one might expect that the carbon-rich aerosol observed in the plume of jet aircraft may be influenced strongly by the aromatic content of the fuel in the same fashion that aromatic hydrocarbon content of gasoline correlates with urban aerosol development (Odum et al., 1997). No published reports confirming this speculation have appeared to our knowledge. This issue should be given attention in future planning of bench tests on aircraft engine emissions.

In addition to uncertainties related to hydrocarbon species, no detailed analysis of the nitrogen-containing compounds in the current jet fuels has yet been carried out to our knowledge. Within the temperature range used for the distillation and separation of the jet aircraft fuel (~170-260°C), compounds such as alkyl-substituted pyridines (azines), pyridazenes (1,2-diazines), pyrimidines (1,3-diazines) and pyrazines (1,4-diazines) would be expected to be present, unless the jet fuel components receive some special chemical treatment to eliminate nitrogen-containing organics in the fuel preparation. The CRC Handbook of Aviation Fuel Properties (CRC, 1984) states that non-hydrocarbon compounds containing sulfur, oxygen, or nitrogen are found in low concentrations in aviation fuels. Even though the amount of nitrogen-containing species in the fuel may be small, the NO_x formed from their combustion may not be a trivial contribution to total NO_x emissions from the aircraft. Each nitrogen atom present in the fuel will appear as an NO_x species in the exhaust, and unlike the NO_x produced by the high-temperature N_2–O_2 combustion of air, this fuel-related NO_x will probably not be altered by the combustion conditions (excess of air, fuel, temperature profile, etc.). It would be worthwhile for SASS to give more consideration to this potential source of NO_x.

Tropospheric Ozone Trends

Current SASS studies are concerned principally with the estimation of upper-tropospheric ozone levels. An important source of upper-tropospheric ozone, however, may be the upward transport of lower-tropospheric ozone. A reasonably accurate lower-tropospheric ozone field is therefore required in models. Unfortunately, tropospheric ozone observations are not extensive today, and their interpretations differ. For instance, on p.40, Friedl (1997) states, ". . . in the lower troposphere, there are indications of ozone increases over the past 25 years in parts of northern mid-latitudes, but the increase appears to have leveled off since the mid-1980s over Europe and the United States." Yet, some of the recent studies cited by Friedl do show increasing trends, as do other observations at northern mid-latitude sites, such as those over Japan (Akimoto et al., 1994; Lee et al., 1998).

In addition, Kley et al. (1994) have pointed out that in some cases the "potential ozone" ($[O_3] + [NO_2]$) of an air mass, rather than $[O_3]$ alone, may be a more appropriate indicator to use. Local emissions of NO titrate existing ozone, forming NO_2, but as transport and dilution of the air mass occurs, NO_2 photolysis will eventually form additional ozone. The SASS project needs to take into account the possible influence of tropospheric ozone increases in the years ahead, which would affect model results and uncertainties.

Simulation of j-Values

j-values describe the extent of photodissociation that occurs for any particular molecule as a function of the incident radiation. The accurate representation of *j*-values in tropospheric chemistry simulations remains a difficult problem. Brasseur et al. (1996) shows that using a diurnal cycle in *j*-value calculations, rather than average *j*-values, has a large effect (>25 %) on the calculated maximum in the NO_x mixing ratio and the net ozone production rates. It thus appears to be necessary to use either the diurnal cycle or a well-calibrated algorithm to adjust the results when average *j*-values are used.

The calculated *j*-values for the cloud-containing troposphere are a matter of great concern to modelers of tropospheric ozone, since the use of clear sky *j*-values in simulations does not reflect the reality of the often cloudy troposphere. Recently, in NASA's SONEX campaign, direct aircraft measurements of the spectral solar flux were made using well-calibrated, 360° spectroradiometers for a variety of locations within the upper troposphere, for both non-cloudy and cloudy conditions (including below-cloud, above-cloud, and in-cloud flight paths). From these measurements, realistic *j*-values can be calculated for the important light-absorbing molecules (e.g., O_3, NO_2, CH_2O) for a variety of typical cloudy tropospheric conditions. The extent of cloud cover over the Earth can be estimated from suitable satellite or other databases. These data, coupled with the extensive and growing database of *j*-values measured for various amounts and types of clouds, should be used to develop algorithms that can estimate realistic *j*-values for use in the predictive modeling of the effects of aircraft emissions on the ozone column.

PARTICLE EMISSIONS, CHARACTERIZATION, AND CHEMISTRY

The SASS strategic plan reflects an increased recognition of the importance of aerosol-related issues, and more of its resources are now being directed toward this area. This greater emphasis reflects primarily the growing awareness of the potential influence on climate of aircraft exhaust particles, through both direct and indirect effects. PAEAN's interim review of SASS (NRC, 1997a) made a number of recommendations related to aerosols, including the following:

• Designate a team of researchers to examine extant data sets (U.S. and other) for the mid-troposphere, to assess the extent to which they provide a consistent picture of the aerosol and gas-phase characteristics of the free troposphere, and its regional variability.

• Use these data sets and other information to bound current uncertainties and sensitivities of the relationships among clouds, aerosol, and radiative effects.

• Evaluate and prioritize research strategies on the basis of these existing data sets and uncertainties.

• Increase efforts to characterize the size and properties of soot particles emitted under ambient operating conditions

The recent SASS and European assessment reports indicate that progress has been made in some of the above areas. However, important details still remain poorly understood. This section addresses several issues related to aerosols and contrails that require more investigation. Among them are the emission, formation, and ambient distribution of sulfate and carbon aerosols, the formation of contrails and their role in cirrus-cloud development, and the heterogeneous chemistry occurring on all of these surfaces.

Sulfur Aerosols

Friedl (1997) notes a major difficulty in quantitatively understanding how sulfur compounds in aircraft fuel are converted to sulfate in the aircraft engine. The sulfur content of jet-aircraft fuels varies with each fuel's specifications, but it is commonly around 200 ppm by weight or lower. As the fuel is burned in the engine, the chemically bound sulfur compounds occurring in the hydrocarbon fuel matrix form SO_2, SO_3, and H_2SO_4 and its hydrates.

An experimental determination of aerosols in aircraft wakes was made in 1995, when an ER-2 aircraft was able to sample the exhaust plume of a Concorde aircraft (Fahey et al., 1995a). A huge number of aerosol particles was found in the plume with peak values ranging up to 15,000 particles/cm^3 (the background concentration was approximately 6-18 particles/cm^3). Heated at 192°C, a large fraction of these submicron particles was volatilized, and their composition was consistent with that of sulfuric acid. While the Concorde engine design is quite different than that of today's subsonic fleet, these studies are still relevant to SASS, as they confirmed that a large number of aerosols can be generated in the aircraft wake by the simultaneous condensation of sulfuric acid and water vapors (heteromolecular nucleation), even if the local atmosphere is undersaturated with water vapor. Similar conclusions were reached by Schlager et al. (1997) when they sampled flight corridors in the troposphere and in the stratosphere, showing the similarity between supersonic and subsonic plumes with regard to aerosol formation.

Several theoretical studies have been performed to predict aerosol production in aircraft wakes, including those of Miake-Lye et al. (1993, 1994), Zhao and Turco (1995), Kärcher (1995), Danilin et al. (1997), Yu and Turco (1997), and Taleb et al. (1997). The mechanism of aerosol production involves the formation of sulfuric acid in the jet regime, followed by its condensation with water vapor (heteromolecular nucleation). The rate of formation of the aerosols is generally calculated according to the "classical" theory of (binary) nucleation. When this theory is applied to aircraft wakes, nucleation rates as high as 10^{12}/cm^3 are predicted, depending on the sulfur content of the fuel. However, it has been shown that the steady-state assumption used in the classical theory is not valid (Taleb et al., 1997), and that a certain delay is needed to reach the steady state, the net effect being a reduction of the number density of the newly formed particles.

Other refinements to nucleation theory include the formation of hydrates in the gas phase, as well as the role of ions (Yu and Turco, 1997). So far, it seems that only binary nucleation on ions can explain the presence of "large" particles (> 9 nm) observed in the wake of the Concorde.

Turco and Yu (1997) pointed out that the total number density of aerosols present in a plume, arising from the coagulation of particles and from the dispersion of the plume, can be predicted to a certain extent without knowing the detailed mechanisms of their formation. This probably applies for time scales of several minutes, but these detailed mechanisms must still be known in order to describe the interaction between soot and aerosols in the early stage of the plume. More complete data on the mode of formation and growth of the aerosols should be pursued if adequate parameterizations of the aerosols' radiative and chemical impacts are to be developed.

The field observations of Fahey et al. (1995a,b) and others suggest that more than 10% of the sulfur appears in the fresh jet-exhaust plume as sulfate (presumably H_2SO_4 and its hydrated forms); yet modeling studies by Brown et al. (1996a), Miake-Lye et al. (1994), and Kärcher et al. (1996), which simulate the chemistry within the engine and in the plume, suggest that only 1-2% of fuel sulfur is converted to SO_3 and H_2SO_4. This discrepancy between measurement and model may be caused in part by the difficulty of modeling the highly complex transport and combustion processes in a jet engine, or by uncertainties in the determination of the particle size distribution.

The extent to which SO_3 (and subsequently H_2SO_4) is formed should be explicable in terms of the complex kinetics of the many reactions involved, but the rate-coefficient data needed for the calculations are often not available for the temperature-pressure regimes occurring in the aircraft engines. Although this "kinetic solution" is the ultimate goal to aid our understanding of sulfur conversion, the thermodynamic properties of the sulfur gases also yield useful information. At the high temperatures occurring in the engine (~1400 K after the combustion chamber and ~600 K at the exhaust), the rates of reactions converting SO_2 to SO_3, and the reverse transformation of SO_3 to SO_2, are fast; however, current calculations suggest that typical combustion residence times are on the order of 1-3 ms, while somewhat longer times (about 5-6 ms) may be required to establish concentrations of those species close to their equilibrium values at the particular local temperature.

The fraction of SO_x that appears as SO_3 when equilibrium is achieved, $2SO_2 + O_2 \leftrightarrow 2SO_3$, depends on the oxygen content of the fuel-air mixture in the engine. It is difficult to know the distribution of oxygen in the gas flow in the various regions of the engine, but it is likely that during much of the transport of gases through the engine, some excess of oxygen over that required for complete oxidation of the fuel is present. If equilibrium between the sulfur oxides and oxygen is maintained over the range of temperatures approaching those found near the output of the engine exhaust—that is, the attainment of equilibrium during passage through the engine is not limited by the kinetics—then the relatively large fractional conversions of SO_2 to SO_3 that have been observed (e.g., by Fahey et al., 1995a,b) are not unexpected. Precise knowledge of this conversion factor is desirable because it controls the aerosol formation rate and the sulfate loading of the atmosphere.

Incomplete reaction-rate data, as well as factors peculiar to the engine design (the effects of engine walls on the combustion, the extent of carbonaceous aerosols present, time of flow through the various chambers in the engine, and the temperature profile of the flowing gases), complicate calculation of the changes occurring in the fuel-air mixture. Such calculations are important, however, for understanding the extent of sulfur conversion to SO_3 and H_2SO_4 in the engine. To carry them out, it may be necessary to determine experimentally the missing rate-coefficient data for the potentially important SO_2 oxidation steps. Some of the reactions that may be important in establishing the conversion of SO_2 to SO_3 in the jet engine are:

(1) $O + SO_2 (+ M) \rightarrow SO_3 (+ M)$
(2) $O + SO_3 (+ M) \rightarrow SO_2 + O_2 (+ M)$
(3) $HO_2 + SO_2 (+ M) \rightarrow OH + SO_3 (+ M)$
(4) $OH + SO_2 (+ M) \rightarrow HOSO_2 (+ M)$
(5) $HOSO_2 + O_2 \rightarrow HO_2 + SO_3$
(6) $SO_3 + H_2O (+ H_2O) \rightarrow H_2SO_4 (+ H_2O)$

Reaction rate coefficients for most of these reactions are poorly known and are difficult to obtain for the range of temperatures and pressures encountered in the jet engine. Yet, these data are required to derive more meaningful theoretical estimates of the extent of sulfur conversion to SO_3 in the engines.

Finally, it should be noted that the relative significance of this aircraft-related particle production must be assessed in terms of other natural processes that also produce high concentrations of new aerosols over extensive spatial scales. High concentrations (several thousand to tens of thousands per cm^3) of recently formed nuclei, associated with cloud outflow, have been observed at altitudes of 8-10 km. Such regions have been reported to extend over hundreds of kilometers, for outflow from deep convection near the Intertropical Convergence Zone (Clarke, 1993). More recently, outflow from lower clouds in mid-latitudes has been shown to produce similarly high particle concentrations in the daytime, by photochemistry linked to high sulfuric acid production near these clouds (Clarke et al., 1998a,b).

Soot Emissions

The SASS interim assessment indicates that considerable progress has been made in understanding various aspects of the emission of soot particles from aircraft. Soot has many characteristics that are not fully understood, however, and they may have important implications for heterogeneous chemical processes and for aerosol nucleation, growth, evolution, and role as cloud condensation nuclei (CCN). Soot can provide a surface for the conversion or deposition of sulfates, which in turn can influence the growth of the aerosol and uptake of water. It will be important to estimate the contribution of water uptake to measured (or inferred) aerosol properties, since in many cases aerosol size and composition reflect the concentration of aqueous solution and frozen droplets in equilibrium at ambient relative humidity. Thus, soot, sulfates, and water must be considered as an integral system in the assessment of aerosol effects.

While measurements of tropospheric aerosol and soot are increasing, the uncertainties related to the impacts of these particles are not necessarily decreasing. The uncertainties persist in part because of the difficulty of measuring and identifying the soot component of aircraft emissions under the appropriate conditions. Some data and suggestions presented in the recent literature are inconclusive and, at times, contradictory. More data will be needed to confirm the existing observations and to determine the significance of aircraft soot emissions relative to other sources of soot. Some recent publications are discussed below to illustrate the complexity of the role soot aerosol may play and the many related uncertainties.

In a 1995 paper, Blake and Kato drew several conclusions that are relevant to aircraft soot emissions: " *(i)* During volcanically quiescent periods, the calculated total surface area of black carbon soot aerosol is of the same order of magnitude as that of the background sulfuric acid aerosol . . . *(ii)* mass balance calculations suggest that aircraft soot injected at altitude does not represent a significant source of condensation nuclei for sulfuric acid aerosols. . . . *(iii)* The measured latitudinal distribution of this soot (from 90°N to 45°S) at 10- to 11-km altitude is found to co-vary with commercial air traffic fuel use, suggesting that aircraft fuel combustion at altitude is the principal source. . . ."

The first conclusion is justified only if the surface area of the soot is calculated on the assumption that the carbon chains are made up of 20-nm isolated spheres that are not collapsed upon themselves. This value may not accurately describe the average condition. Similarly, evidence suggests that a significant amount of the soot surface becomes coated by other species as the plume evolves; yet, the second conclusion is based on a technique that cannot detect the presence of soot inside the sulfate particle. Finally, the third conclusion is based on a very limited dataset collected over a small altitude range and does not account for the fact that the observed soot aloft also co-varies with major surface sources of continental aerosol advected from Asia at altitudes as high as 12 km over the Pacific (Menzies and Tratt, 1997). While Blake and Kato deduce from their uniformly low measurements of upper-troposphere tropical soot that biomass burning does not constitute a significant soot-aerosol source at altitude in the tropics, growing bodies of data (e.g., from PEM-Tropics) suggest that those low measurements may not be adequately representative; pollution layers are often vertically stratified, and thus may not be representatively sampled if measurements are made over a narrow altitudinal range.

Two interesting conclusions are drawn in a recent paper by Pueschel et al. (1997). The first is that a strong gradient in black carbon aerosol (BCA) exists between the northern and southern hemispheres, indicating mixing times longer than lower-stratospheric residence times. The second conclusion is that BCA is generally observed to 20 km altitude, so if subsonic commercial aircraft are the major source of lower-stratospheric BCA, a mechanism must exist that transports BCA from flight levels below 12 km up to 20 km. The possibility exists that a significant fraction of the soot observed near tropopause levels may arise from biomass burning, and may be quite unrelated to aviation. High concentrations of carbon aerosols (both black carbon and organic) have been observed to be convected over extensive regions of the tropics (Andreae and Crutzen, 1997). In the 1995 NASA PEM-Tropics experiment, biomass-burning plumes were observed at 8-10 km altitude over the remote South Pacific, clearly revealing that strong biomass-burning sources exist. Measurement of specific markers to help isolate and identify these "alternate" soot sources (e.g., CO/ethyne ratios, which are quite different for emissions from biomass burning and from aircraft) should be incorporated into future sampling strategies.

Interpreting the limited measurements of soot currently available is difficult, especially when their sparseness is coupled with recognized sampling uncertainties (Penner and Novakov, 1996). While it is clear that aircraft do introduce soot aerosol into the free troposphere, the contribution of their emissions relative to that from other sources remains highly uncertain. Reducing this uncertainty, and putting aircraft measurements of soot in the context of this uncertainty, should remain a priority of the SASS program.

Aerosol Size Distribution

"Measured concentrations of volatile particles in commercial aircraft wakes are large and show significant unexplained variability." This statement, or something like it, has appeared repeatedly in various papers and documents, with particular reference to the high particle-number concentrations observed behind the Concorde by Fahey et al. (1995a,b). Many of these studies make assumptions about the aerosol mass (presumably related to carbon and sulfur emissions in the fuel) associated with the number concentration of particles observed. In most cases, some estimate of the relationship of mass to number is made, but the uncertainties in these assumptions are often poorly established and are often not properly propagated through the calculations. To its credit, the Friedl (1997) report has provided (pp. 34-35) loose bounds on this assessment that suggest a globally averaged enrichment of CCN spanning two orders of magnitude, from 0.8 to 77 cm^{-3}; these values, however, range from insignificant to probably important. We agree with that report's conclusion that more focused research is needed to reduce these uncertainties.

It should be emphasized that the key to reducing these uncertainties will be careful characterization of the aerosol size distribution, from about 3 nm up to 3000 nm or larger. This precision is necessary because a few large particles can dominate aerosol mass in the presence of smaller particles, while the number of smaller particles may actually exceed the large particles in number by several orders of magnitude. For aerosol/cloud issues, it is generally the aerosol number that determines cloud droplet number and the associated radiative effects. For interpretation of specific chemical conversion rates or fuel emission scenarios, it is the aerosol mass that is important. Last, the available aerosol surface area, which is often dominated by aerosols of intermediate sizes, determines heterogeneous chemical properties that are important to a multitude of processes, including the growth and evolution of aerosols.

The issue of mass and number interpretations comes up in the comparison of emission inventories for soot in the SASS assessment. On page 32 of Freidl (1997) it is argued that "in the worst case, aircraft are only responsible for <0.5% of the total global soot emissions." Even if this figure is correct on a mass basis, most emission indices for soot are obtained for combustors that generally emit larger carbon-soot particles than aircraft engines and, thus, fewer particles for a given amount of soot mass. Hence, the number of soot particles emitted by aircraft engines could be much larger than the 0.5% mentioned. Similarly, in discussion of the possible role of soot as a condensation site for sulfuric acid and its potential contribution to CCN, we are most interested in the number concentration of soot particles. Any comparisons of aircraft soot-emission indices with those of other sources can be misleading if both are mass-based when aerosol number is the concern.

The interpretation of fundamental processes related to the production of aerosol from aircraft requires that the entire aerosol size spectrum be fully characterized and its dynamic evolution understood. The integral properties of the size distribution (total number, surface, volume) are of limited use and provide no opportunity to evaluate the evolution of the size distribution. Consequently, it is important that future measurements and model assessments include adequate determinations of, or realistic parameterizations of, the complete size spectrum.

Need for an Aerosol Climatology

At present, the effects of tropospheric aerosols are the largest uncertainties in quantifying climate forcing due to anthropogenic changes in the composition of the atmosphere. The main reason for these uncertainties is a lack of understanding of the contribution of the natural background aerosol to the total particle burden of the troposphere, which makes it difficult to evaluate the relative importance of any additions. The new SASS strategic plan calls for building an aerosol climatology database. Such a database will be fundamental to putting any assessment of aviation-related aerosol radiative effects or influences on tropospheric chemistry in perspective.

Providing a representative aerosol climatological database for the free troposphere will be a significant effort. It will have to build on careful assimilation of data from as many aircraft datasets as possible (NASA-GLOBE, NASA-PEM Tropics; NASA-TRACE, ACE-1, ACE-2, and others), which are large and diverse and will take time to assemble. At the same time, they will probably not provide full global and temporal coverage. In situ measurements can fail to capture the complex aerosol distribution (which often occurs as layers or 'rivers' of aerosol). Hence, whenever possible, such datasets will need to be integrated with satellite and lidar data in order to improve our understanding of the structure of aerosol fields in the troposphere. Because significant amounts of surface aerosol can at times be carried aloft into transport paths that coincide with major air corridors, the interpretation of differences in aerosol type found inside and outside of air corridors will require care.

PAEAN recognizes that the establishment of a representative aerosol climatology may be beyond the scope of SASS's current mandate and resources; however, studies currently active under NASA's OES (in particular, through the Global Aerosol Climatology Project), as well as other programs outside of NASA, already include measurement efforts that can contribute directly to the needed climatology. Closer collaboration with these programs would be highly desirable.

The panel also recognizes that compiling datasets from different sources is not a straightforward process. Other programs may not be measuring all the parameters that are most important to SASS's analyses, and the complicating factors raised in the previous section (such as particle mass vs. number) must be carefully considered with any aerosol datasets. Yet, it seems well worth the effort to try to meet such challenges. This collaborative approach may be the only currently available means of collecting the data necessary to fully evaluate the relative importance of aircraft particle emissions.

Contrail Formation

Contrails may be formed entirely from direct emission of precursor substances, or may result from the emissions of smaller particles that modify or enhance natural cirrus. Compared to natural cirrus clouds, crystals in aged contrails have a greater number density and smaller size (Gayet et al., 1996). There is a wide range of independent evidence that contrails occur both alone and in combination with natural cirrus clouds and that their spatial extent may reach synoptic scales over many regions of Earth (Sassen, 1997; Sausen et al., 1997). Contrails interact with natural cirrus to stabilize the cirrus, slowing its sedimentation and thus increasing its lifetime.

Unlike acidic aerosols, which are always found in aircraft wakes, contrails form only when the ambient meteorological conditions are favorable. Contrails appear when the hot water vapor emitted by the engines mixes with moist ambient air. Appleman (1953) derived a criterion for contrail formation, assuming that during the mixing process water vapor must reach saturation. This condition restricts the formation of contrails to the troposphere, although they have been observed in the stratosphere under very cold conditions. Appleman's criterion has often been used to predict threshold conditions of contrail formation; it is unable to predict their lifetime though, because it does not take into account the kinetics of particle growth or evaporation.

Discrepancies have been found between observed and computed threshold conditions for contrail formation (Busen and Schumann, 1995), indicating that the details of the formation mechanisms are not fully understood. The main question is which nucleation mechanism is activated. Since the saturation ratio of water vapor is not large enough to induce homogeneous nucleation of water vapor, heterogeneous nucleation on existing particles must take place. Several pathways of contrail formation have been considered, including heterogeneous nucleation of water vapor on frozen H_2SO_4/H_2O aerosols, on soot, and on soot coated with H_2SO_4. This last case can occur either by adsorption of H_2SO_4 from the gas phase or by coagulation of soot particles and supercooled H_2SO_4-H_2O aerosols. In addition, observations from the SUCCESS campaign indicate that ambient aerosols, including mineral particles, may play an important role in contrail formation (Twohy and Gandrud, 1998; Jensen et al., 1998). Studies by Busen et al. (1998) indicate that fuel sulfur content (FSC) has little effect on the formation of visible contrails or on the threshold conditions for contrail formation. Further studies of contrail formation mechanisms would be worthwhile, as this uncertainty limits our ability to assess contrail chemical and radiative impacts.

Heterogeneous Chemistry

Participants in AEAP and the European assessment programs recognize the importance of including heterogeneous chemistry in their modeling efforts. However, implementing chemical modules that incorporate the possible reactions is difficult and introduces significant uncertainties. Only a few heterogeneous reactions, occurring on polar stratospheric cloud (PSC) particles and sulfuric-acid aerosols, have been identified and sufficiently quantified to allow inclusion in both assessment programs. Discussed below are the heterogeneous processes that may be of particular importance to subsonic aircraft impacts on atmospheric chemistry.

N_2O_5-Aerosol Reactions

The importance of including in models the reactions of N_2O_5 on moist aerosols [N_2O_5 + H_2O (in aerosols) → $2HNO_3$] is clear from the studies of Dentener and Crutzen (1993). Their tropospheric simulations, which include reaction of N_2O_5 with aqueous aerosols (using an uptake coefficient of 0.1), show yearly global average decreases in the concentration of tropospheric NO_x of as much as 50% and of O_3 by 9%. These reactions are most important during the winter months. For instance, the calculations of Dentener and Crutzen (1993) indicate that, for January in the troposphere north of 45° N, more than 90% of all NO_x is removed by this heterogeneous reaction.

These results are relatively insensitive to the assumed value of the uptake coefficient. All experimental measurements are in relatively good agreement on the magnitude of the uptake coefficient for aqueous sulfuric acid aerosols, and the dependence of this value on temperature and relative humidity (Fried et al., 1994). Experimental estimates of the uptake coefficient for aqueous ammonium sulfate and ammonium bisulfate aerosols, common to the lower troposphere, show some scatter, but all lie within the range 0.02-0.1 (Hu et al., 1997).

Both the U.S. and European scientists recognize the importance of the N_2O_5 + H_2O reaction, and it is included in some current global atmospheric models. Probably the biggest problem in including this reaction in models lies in estimating the magnitude of the moist aerosol surface area and its geographical distribution and seasonal variation. This again emphasizes a need for better aerosol data and a plausible aerosol climatology if these uncertainties are to be reduced.

Removal of HO₂ and OH Radicals in Heterogeneous Reactions

Other potentially important heterogeneous reactions involve two key participants in atmospheric chemistry, the HO_2 and OH radicals. Laboratory experiments show that reactions that remove each of these species on sulfuric acid, ammonium sulfate, and other aerosols have significant mass accommodation coefficients (> 0.2). Such reactions could provide a source of highly reactive species that promote solution phase reactions in aerosols or in cloud water, and they can act as a significant sink for HO_2 and OH. These important chain-carrying radicals are key participants with hydrocarbons and NO_x in promoting ozone generation in the troposphere (Mozurkewich et al., 1987; Hanson et al., 1992; Cooper and Abbatt, 1996).

It is stated on p. 38 of Friedl (1997) that the HO_2 and OH reactions have not yet been well enough characterized to be included in models of the upper troposphere. However, this lack of characterization should not exclude their use in model sensitivity tests of the effects of aircraft emissions on upper tropospheric and lower stratospheric ozone. A series of sensitivity tests should be made with current models in which these reactions are included with a range of aqueous aerosol concentrations and uptake coefficients that cover the range of experimental values. The absence of these reactions from the current assessment models should be factored into estimates of the overall uncertainties by the NASA and European assessment studies.

Reactions on Soot Particles

O_3, NO_2, HNO_3 and other good oxidizing agents, in theory, can react readily with black carbon aerosols. However the ozone-carbon reaction is relatively slow for upper tropospheric temperatures (Stephens et al., 1989), and the oxidation of the surface layer strongly inhibits subsequent reaction. Lary et al. (1997) point out that the reduction of HNO_3, NO_2, and O_3 on carbon aerosols may be an important effect of increased air traffic that has not been considered to date. If these are important, then a significant lower stratospheric ozone loss mechanism could exist. The amount of carbon aerosol emitted by aircraft engines has been estimated experimentally, but it is currently not known what fraction of the aerosol is quickly coated by H_2SO_4 or other adsorbed material, which would lower the available free-carbon surface area of the aerosols.

Kärcher (1997) has considered the possible role of aircraft exhaust soot in promoting heterogeneous reactions (involving HNO_2, SO_2, and NO_2) in aircraft plumes. He suggests that rapid heterogeneous reaction of exhaust NO_2 with soot might explain why the observed NO_2 values in the Concorde plume were a factor of 2 lower than indicated by a photochemical steady-state approximation, and considers it likely that soot particles absorb oxidized sulfur gases at emission and collect volatile H_2SO_4/H_2O in the plume. If the aerosol is covered with sulfuric acid or other material that can absorb water from the air, the resulting aqueous solution can promote reactions at the solid-liquid interface. Active sites can be regenerated if the reaction product is soluble in water and leaves the aerosol surface. As mentioned earlier, it is not yet clear whether the abundance of carbon aerosols in the upper troposphere and lower stratosphere is significantly increased by aircraft exhaust emissions. If this is indeed the case, it is important to determine experimentally the possible extent of alteration of the ozone column due to heterogeneous reactions involving those aerosols (whether coated with H_2SO_4 or uncoated).

Reactions Between Halogenated Species on Cirrus Clouds

Raman lidar measurements of ozone, water vapor, and cirrus cloud optical properties over northern Germany (Reichardt et al., 1996) showed decreased ozone levels in the upper troposphere in the presence of ice-cloud layers. Borrmann et al. (1997) observed elevated ClO levels in and near cirrus clouds near the tropopause; they suggested that heterogeneous chemistry might be generating active chlorine species, ClO and Cl, from relatively inactive $ClONO_2$ and HCl (and also HOCl with HCl). Similar reactions have been observed in the arctic "ozone hole" (HCl + $ClONO_2$ → Cl_2 + HNO_3). Upon dissociation by sunlight, Cl_2 can initiate the catalytic cycles that destroy O_3.

From their modeling studies, Solomon et al. (1997) conclude that cirrus clouds, occurring with sufficient frequency and spatial extent, could influence chemical composition and ozone depletion in the region near the tropopause. While the presence of cirrus clouds near the tropopause plays only a small role in determining the total ozone column trends (less than 1.5% change in computed column ozone at mid-latitudes in these calculations of Solomon et al.), they are of particular importance in determining the changes in ozone at the tropopause and hence the radiative forcing. Solomon et al. note that emissions from aircraft (contrails and chemical effluents) could influence cirrus cloud distributions and frequency. Clearly, if there were to be variability or trends in the frequency of occurrence of cirrus clouds or in their distribution, this could add substantially to their impact on the ozone layer. As our knowledge of the influence of aircraft emissions on contrail formation and the possible enhancement of cirrus cloud formation improves, the resulting effects on the ozone column should be tested in the modeling studies.

ATMOSPHERIC DYNAMICAL PROCESSES

Atmospheric dynamics, operating over a vast range of spatial scales, includes many process that must be understood in order to properly assess the impacts of aircraft exhaust. In particular, since aircraft emissions are deposited primarily in the upper troposphere and lower stratosphere, cross-tropopause transport processes may significantly affect aviation's impacts in both regions. Stratosphere-troposphere exchange directly affects the distribution of aircraft emissions and indirectly affects the chemical impact of these emissions by influencing the composition of the background atmosphere.

At low latitudes, gas and particulates are convectively transported from the troposphere upward to the stratosphere. Downward transport occurs at higher latitudes through tropopause folding, and isentropic transport occurs through subtropical tropopause breaks. There are indications of distinct hemispheric asymmetries and seasonal variations in these cross-tropopause processes.

In the last few years, our understanding of the large-scale transport processes has improved significantly. These developments are discussed in detail in Holton et al. (1995), as well as in the panel's recent interim review of AESA (NRC, 1998). However, the smaller-scale details of cross-tropopause transport are still not well enough known to permit appropriate representation in models. Additional balloon and/or aircraft campaigns designed to probe the tropical and subtropical tropopause regions, and measurements of a variety of long-lived tracers in these regions may provide the data needed to help understand these transport processes.

This issue needs to be an important focus of AEAP analyses, but little mention is made of it in the SASS strategic plan. The issue is relevant to both the subsonic and the supersonic components of AEAP; thus it is important that SASS and AESA sort out this common concern with whatever type of collaboration will be most effective.

CLIMATE IMPACTS

Gases and particles emitted by jet aircraft can affect climate in numerous ways, through both direct and indirect radiative forcing. Some of the most uncertain and potentially important of these processes are discussed below. Discussion of climate modeling is reserved for the following chapter.

Effects of Radiatively Active Gases

As discussed in the NASA assessment (Friedl, 1997), it may turn out to be very difficult to quantify the climatic effects of directly emitted combustion gases (such as CO_2 and water vapor) through observational studies; however, provided the magnitude of the emissions can be reliably predicted, there is some hope that their effects can be estimated with reasonable accuracy through model computations.

A more indirect and uncertain problem is that some combustion gases such as NO_x can change ozone chemistry in the upper troposphere/lower stratosphere region, which in turn can have significant climatic impacts. Estimating the magnitude of these impacts requires keeping at least three factors in mind: (i) ozone loss in the upper troposphere/lower stratosphere region may be linked to heterogeneous chemistry on cirrus cloud particles (Borrman et al., 1997); (ii) although the ozone changes in this region may have a minor influence on the total ozone column, such changes may be quite important to radiative forcing of the climate system (Solomon et al., 1997); and (iii) there are possible feedbacks through ozone loss/temperature decrease/cirrus formation to consider. Understanding the ozone changes in this altitude range will be critical to assessing climate forcing and response; however, modeling or observing such perturbations in the region around the tropopause is very difficult because large variations occur in both transport patterns and concentrations of key species.

Effects of Particles and Contrails

The direct radiative forcing caused by jet aircraft soot and sulfur emissions is thought to be relatively insignificant. However, contrails and injected aerosols can modify the abundance and microphysical properties of cirrus clouds, which in turn can have significant climatic impacts. Yet, these cloud-related effects are very difficult to quantify because of the large uncertainties about ice-crystal formation mechanisms and physical properties. To model the radiative effects of contrails and injected aerosols, one needs to know particle shape, size distribution, and refractive index. During the recent SUCCESS campaign, a great deal was learned about ice crystal size and shape characteristics (for example, see Lawson et al., 1998; Goodman et al., 1998; Liou et al., 1998; Sassen and Hsueh, 1998), and about the role of ice nuclei in cirrus ice formation (Chen et al., 1998;

Rogers et al., 1998, De Mott et al., 1998). Yet, even if these physical properties were fully characterized, it is quite difficult to compute the optical properties of such varied particles.

The growth of contrails into extensive, diffuse "contrail-cirrus" was clearly documented by geostationary satellite observations during the SUCCESS mission (Minnis et al., 1998). Many earlier investigators had noted this link between aircraft activity and enhanced cirrus-cloud coverage (for example, Machta and Carpenter, 1971; Chagnon, 1981. It has been suggested that contrails are responsible for a 2% increase of cloudiness above the United States between the years 1950 and 1988 (Angell, 1990); similar findings were also reported by Liou et al. (1991) above Salt Lake City, and by Bakan et al. (1994) over Europe.

Since the physical properties of contrails appears to be quite different from naturally occurring cirrus clouds, the mechanisms that link contrails to climate may be very different from those operating in the natural, unperturbed system. The balance between the albedo and greenhouse effects of contrails determines the net radiative forcing, and this may be a function of a wide range of environmental factors. The uncertainty in the net forcing is substantial and may be of opposite sign in different atmospheric regimes. The chemistry and physics underlying this complex issue are discussed in many of the SUCCESS papers cited above as well as several other recent articles (see, e.g., Andronache and Chameides, 1997; Baker, 1997; Sassen, 1997; Szyrmer and Zawadzki, 1997; Travis et al., 1997).

Although the current understanding of direct and indirect effects of aircraft particle injections is discussed at some length in the NASA report, a cohesive and well-articulated plan for how to proceed is generally lacking. In view of the complexity of the problem, this is perhaps not surprising. Some effort should be directed toward exploiting the possibility of using new remote-sensing capabilities to study these aerosol/cloud/contrail issues; any such studies should be validated through comparison to in situ measurements whenever possible. In the near term however, given the complexity of the problem and the limited resources available through the SASS program, back-of-the-envelope calculations may have to provide the guidance for how to proceed. These should be used to put plausible bounds on potential effects and to point to areas of sensitivity. In conjunction with emerging data from SUCCESS and other recent large programs (ACE-1, PEM, etc.) efforts should be made to gauge aircraft perturbations against other major natural and anthropogenic inputs (biomass burning, urban pollution) that can impact tropospheric aerosol size, concentration, composition, and cloud nucleating capability. An understanding of both absolute and relative effects caused by aerosol emissions from aircraft will need to be developed further.

3

Modeling Considerations

The accuracy of current calculations of aviation's impact on the atmosphere is restricted by insufficiencies in atmospheric data and inadequate representation of some key physical and chemical processes in models. The lack of a suitable measurement database currently prevents full testing of models by comparing their results with atmospheric observations. At present, no model can treat in detail all the important processes, which operate over orders of magnitude in spatial and temporal scales. Thus, a hierarchy of models, from box models to global-scale 3-D models, is being used to evaluate the impact of aircraft emissions on the atmosphere. Several types of models were used for the NASA assessment (Friedl, 1997) and the European assessment (Brasseur et al., 1997), as discussed below. Note that none of these models are presently able to effectively incorporate heterogeneous processes on aerosols or aerosol/cloud interactions.

STUDIES WITH EXISTING CHEMICAL-TRANSPORT MODELS

Six chemical mechanisms were evaluated for the SASS interim assessment. All the models were initialized with identical upper-tropospheric chemical and meteorological conditions. Among the mechanisms considered, calculated rates of ozone formation from NO_x-catalyzed reactions in the upper troposphere agreed to within 2-15%. The differences among the models seemed to be related to discrepancies in photolysis rates, nitrogen speciation, and free-radical concentrations.

The ability of the global models used for the SASS assessment to represent rapid vertical transport has been evaluated by simulating the transport of radon, and comparing the results to those obtained by non-SASS models. This comparison has shown that the quality of the assessment models is comparable to that of other chemical transport models, but also that major uncertainties still remain in current model representations of tropospheric processes, such as convective exchanges in the troposphere. An NO_y-tracer study has been used to evaluate large-scale transport in the upper troposphere. As noted in Freidl (1997), the results showed that a vertical resolution better than 1 km at the tropopause level is required to simulate properly the vertical distribution of emissions. When the results of the models used for the NASA assessment are compared with the limited number of available observations, it is apparent that those models all have difficulties in representing some of the features present in the observations.

The European assessment report discusses the results of the GIM (Global Interpretation and Modeling) project of the IGAC (International Global Atmospheric Chemistry) Program, in which the distribution of NO_x was calculated by ten different models. Large differences in NO_x distribution were obtained, reflecting the models' differing transport formulations and source strengths. Most of the models seemed to underestimate the abundance of NO_x and overestimate the HNO_3/NO_x ratio in most regions. This suggests an incomplete understanding of several mechanisms—for instance, heterogeneous chemistry on aerosol particles, or the role of peroxyacetyl nitrate (PAN) in the tropopause region. Our limited knowledge of the NO_x distribution in the free and upper troposphere makes it difficult to evaluate the accuracy of the models' simulated ozone production. Brasseur et al. (1997) also note that the models appear to represent rather accurately ozone distributions in remote tropical and mid-latitude regions, but they underestimate ozone concentrations in other regions, perhaps as a result of poor representation of troposphere/stratosphere exchanges.

The simulation results reported in both documents agree with prior studies in suggesting that aircraft emissions could result in an increase in concentrations of both nitrogen oxides and ozone in the troposphere. The aviation-related NO_x increase in the upper troposphere in the latitude band 30 to 60°N is estimated to be as high as 50%, with ozone increases of a few to 10 pptv (Brasseur et al., 1997). These current estimates of the potential impact of subsonic aircraft are smaller than the estimates of a few years ago, both because current estimates of the emissions are lower and because 3-D models tend to calculate somewhat smaller changes than 1-D or 2-D models.

THE GLOBAL MODELING INITIATIVE

The AEAP's Global Modeling Initiative (GMI) is expected to provide evaluations of the atmospheric impact of the subsonic fleet for future AEAP assessments. AEAP seems to have developed a relatively detailed plan for GMI development and to be making progress in implementing much of this plan. However, before the GMI participates in any assessment exercises, it needs to be carefully evaluated; its results must be compared with those of other models and with available observational data. We note that the stratospheric-related modules of the GMI have participated in international model comparison studies; the tropospheric components of the GMI need to participate in such studies as well.

Intercomparisons involving the GMI should also include a full program of detailed sensitivity studies. Since the results of such studies tend to be highly model dependent, these exercises should involve a range of models beyond the GMI, and should be carefully coordinated. Sensitivity studies should be performed both with the GMI and with existing 3-D chemical-transport models to better evaluate the importance of certain processes that are still not well understood. Among them are:

- convective exchanges and cross-tropopause transport;
- particles in the troposphere and lower stratosphere and possible heterogeneous reactions; and
- partitioning within the different nitrogen species, and the importance of nitrogen reservoir species.

One issue of particular concern to the panel is that the modules being developed under the GMI do not yet incorporate aerosols and their effects, and that it appears unlikely that the GMI will be able to do so realistically in the next few years. If, during the lifetime of SASS, the GMI cannot be employed to address the heterogeneous processes related to aerosols, contrails, and clouds, the large commitment to the GMI on the part of SASS must be questioned. It may be that alternative, perhaps less complex, models that do include aerosols might be more appropriate. Given SASS's required assessment milestone in 2001, it appears urgent to identify the model type and the modeling approach that can best incorporate aerosol effects. A reallocation of some of the resources earmarked for the GMI should be considered, to permit such an effort to be undertaken immediately.

CLIMATE MODELING

Very few global 3-D modeling studies of aviation's climatic effects have been attempted. Adding or removing ozone and/or water vapor in different vertical and horizontal regions in a GCM permit some assessment of the radiative impacts of these gas-phase perturbations. The inclusion of particles in the 3-D climate model is much more difficult, however. As noted earlier, the radiative effects of contrail particles may be very different from those of naturally occurring cirrus, because of the different mechanisms by which they are formed. Contributing to the difficulties of representing the effects of aerosols and contrails in 3-D climate models are the uncertainties in portraying natural clouds in GCMs. Likewise, the observation that aircraft emissions can "activate" cirrus formation needs to be investigated more fully, and a methodology to include this effect in climate models needs to be assessed. The present techniques used to represent clouds in GCMs are highly parameterized; this, in combination with the lack of a detailed treatment of the physics of convection, leads to highly model-dependent results.

Ponater et al. (1996) used a global GCM to examine the climate response to aviation-related water vapor and contrails. They found that the direct radiative perturbations due to water vapor were very small, but that contrails could have a significant effect on climate, with the strongest response occurring under mid-latitude summer conditions; however, a number of uncertainties involving cloud radiative interactions make the results somewhat preliminary. These uncertainties make it much more difficult to use standard 3-D climate models (such as those used to evaluate the climate response to increased atmospheric CO_2, CH_4, N_2O, and CFC concentrations) to assess the climate impacts of contrails.

It is not clear how the climate modeling efforts currently supported by SASS are linked to the other components of the SASS project. Specifically, the climate modeling studies being carried out at NASA-GISS need to be more closely coordinated with the GMI efforts. More generally, SASS needs to encourage its various modeling and experimental groups to compile and evaluate the hemispheric-scale physical and chemical perturbations from aviation with the specific aim of providing databases that are directly useful to climate modelers.

4

Priority Concerns and Recommendations

Over the course of the last few years, SASS-sponsored research has unquestionably made important progress towards its goal of quantifying the atmospheric impacts of aviation, and has also contributed a great deal to our fundamental understanding of atmospheric chemistry and dynamics. Some notable accomplishments include:

- development and testing of several core components of the GMI
- successful completion of the major field campaigns SUCCESS and SONEX
- compiling climatology data on atmospheric reactive nitrogen
- laboratory studies of some key gas-phase and heterogeneous reactions
- development of aircraft fuel-use and emission databases
- compiling a historical database for contrail effects
- improving our understanding of emissions through bench and field tests and plume/wake models

The panel commends SASS on the work done thus far. The panel also commends SASS for responding to both program management recommendations made in the panel's interim review report (i.e., drawing up a detailed strategic plan for the program and putting strong program leadership in place). The results can be seen in the last year's progress.

Throughout the discussions of the previous chapter, many suggestions are made for exploring new issues, or for continuing research activities currently under way. The panel realizes, however, that in the near term SASS may not have the resources to carry out all the desired research activities. Given below then, is a list of issues that the panel would like to highlight as its highest priority. Some of these issues are already alluded to in the SASS strategic plan, but it is not clear that they are given sufficient priority or resources. In addition, suggestions are made about the general strategy and management of SASS. Some of these science and management recommendations echo those made in the panel's interim review of SASS (NRC, 1997a).

SCIENCE ISSUES

Impacts of Heterogeneous Chemistry on Ozone

The impact of aviation on homogeneous ozone chemistry in the troposphere appears to be understood well enough to satisfy SASS's program objectives. However, heterogeneous reactions couple strongly with these gas-phase processes, and may significantly affect ozone concentrations. Thus assessments of the impacts of aviation on ozone levels must include the uncertainties associated with the climatology of aerosols and the heterogeneous processes occurring on these aerosols.

Emissions Characterization

The Freidl (1997) report states that "additional characterization of engine exhaust, particularly of SO_x oxidation products and of particle emissions, is required to define the initial state of aircraft exhuast emissions." The panel agrees that more emissions characterization studies for the species listed above, as well as for a wider range of NMHCs, would be worthwhile. To accurately interpret field measurements of these exhaust species, methods need to be developed to distinguish in situ between emissions from aircraft and those from other sources.

Aerosol Climatology

The chemical and radiative effects of aerosols, contrails, and cirrus clouds remain an unresolved issue, though the panel notes SASS's increased attention to this matter. Aircraft particulate emissions must be assessed in the context of the ambient particle field, however, and it is not clear how SASS expects to accomplish this. The panel recommends that SASS assemble baseline data on the ambient particle fields in the upper troposphere and lower stratosphere, so that the relative effects of aircraft-emitted particles can be properly gauged. SASS should begin by undertaking the collection of currently available aerosol datasets.

Troposphere-Stratosphere Exchange

The upper troposphere and lower stratosphere are highly interactive regions, and cross-tropopause transport may significantly affect aviation's impacts in both regions, since the species emitted by subsonic aircraft are deposited primarily in the tropopause region. Cross-tropopause transport processes are not well enough known to permit appropriate representation in models, and thus need to be an important focus of AEAP analyses. It is important that the SASS and AESA components of AEAP sort out this composite assessment with whatever type of collaboration will be most effective.

The Global Modeling Initiative

Assessments of the atmospheric effects of aircraft emissions come largely from model predictions. Thus, the dependence of the model results upon the parameterizations of sub-grid processes, stratosphere–troposphere exchanges, and the chemical schemes selected needs to be carefully tested. SASS should place a high priority on carrying out detailed sensitivity studies, and the tropospheric components of the GMI should participate in international model intercomparison studies.

A large portion of SASS's current budget is devoted to the Global Modeling Initiative. This level of support will be justified only if the GMI can incorporate aerosol/contrail/cirrus processes into its model framework. Given the complexity of the GMI, in the near term it might be more cost effective to invest in alternative, simpler modeling tools that could include parameterizations of these processes.

Finally, the GMI efforts should be more closely coordinated with the SASS-sponsored climate modeling studies being done at NASA-GISS.

GENERAL MANAGEMENT ISSUES

Planning of Field Campaigns

SASS-sponsored field campaigns like SONEX and SUCCESS have proven extremely valuable in addressing questions related to aircraft emissions, ambient aerosol concentration, and the properties of contrails and cirrus clouds. Given the infrequency with which such complex and costly studies can be undertaken, it is important that they be carefully planned to provide answers to the most pressing uncertainties related to the impacts of aviation. The results of SUCCESS and SONEX should be used to guide the planning and conduct of SASS's next campaign, helping to pinpoint the most important scientific questions and the most effective measurement methodologies.

Cooperation With Other Research Programs

Much of SASS's work thus far has focused on quantifying the absolute magnitude of aircraft emissions and the resulting impacts upon the concentration of atmospheric species such as ozone. We agree that SASS experimental research should focus as much as possible directly on the effects of aircraft emissions. However, the importance of these perturbations can be evaluated only in the context of the natural distribution and variability of the relevant chemical species, as well as the magnitude of other anthropogenic perturbations. Obviously SASS does not have the resources to study the global climatologies and sources for all relevant gases and aerosols (as well as clouds) on its own. SASS should give much higher priority then to exploring how data accumulated by other programs can be used to provide some context in which aircraft effects can be evaluated. We recommend that SASS devote more of its time and money to establishing cooperative efforts with other research programs, both within and beyond NASA, and to evaluating the possible applicability of datasets and information gathered by others. It might even be appropriate to designate an investigator to focus on that task alone.

In particular, some of the research sponsored by NASA's Office of Earth Science may be quite useful in reducing uncertainties, as would some of the climatology database work being coordinated by IGAC. Likewise, SASS should be preparing to use the existing and anticipated data from satellite platforms, identifying the specific data products that will be relevant to evaluating the impacts of aircraft emissions. Methods, infrastructure, and techniques to simulate data flows should be created now, so that when remote-sensing platforms come on line, SASS will be prepared to use the data effectively.

References

Akimoto, H., H. Nakane, and Y. Matsumoto. 1994. The chemistry of oxidant generation: Tropospheric ozone increase in Japan. In The Chemistry of the Atmosphere: Its Impact on Global Change. J.G. Calvert, ed. Blackwell Scientific Publications, Oxford, U.K, pp. 261-273.

Andreae, M.O., and P.J. Crutzen. 1997. Atmospheric aerosols: Biogeochemical sources and role in atmospheric chemistry. Science 276, 1052.

Andronache, C., and W.L. Chameides. 1997. Interactions between sulfur and soot emissions from aircraft and their role in contrail formation. J. Geophys. Res. 102, 21443-21451.

Angell, J.K. 1990. Variation in United States cloudiness and sunshine duration between 1950 and the drought year of 1988. J. Climate 3, 296-308.

Appleman, H. 1953. The formation of exhaust condensation trails by jet aircraft. Bull. Am. Meteorol. Soc. 34, 14-20.

Atkinson, R., D.L. Bausch, R.A. Cox, R.F. Hampson, Jr., J.A. Kerr, M. J. Rossi, and J. Troe. 1997. Evaluated kinetic, photochemical and heterogeneous data for atmospheric chemistry. Supplement V. IUPAC subcommittee on gas kinetic data evaluation for atmospheric chemistry. J. Phys. Chem. Ref. Data 26, 521-1011.

Bakan, S., M. Betancor, V. Gayler, and H. Grassl. 1994. Contrail frequency over Europe from NOAA satellite images. Ann. Geophys. 12, 962-968.

Baker, M.B. 1997. Cloud microphysics and climate. Science 276, 1072-1078.

Blake, D.F., and K. Kato. 1995. Latitudinal distribution of black carbon soot in the upper troposphere and lower stratosphere. J. Geophys. Res. 100, 7195-7202.

Borrmann S., S. Solomon, and D. Baumgardner. 1997. On the occurrence of ClO in cirrus clouds and volcanic aerosol in the tropopause region. Geophys. Res. Lett. 24, 2011-2014.

Brasseur, G.P., J.-F. Müller, and C. Granier. 1996. Atmospheric impact of NO_x emissions by subsonic aircraft: A three dimensional study. J. Geophys. Res. 101, 1423-1428.

Brasseur, G.P., R.A. Cox, D. Hauglustaine, I. Isaksen, J. Lelieveld, D.H. Lister, R. Sausen, U. Schumann, A. Wahner, and P. Wiesen. 1997. European Scientific Assessment of the Atmospheric Effect of Aircraft Emissions. Report to the European Commission (DGXII/D1), Brussels.

Brown, R.C., M R. Anderson, R.C. Miake-Lye, and C.E. Kolb, A.A. Sorokin, Y.Y. Buriko. 1996a. Aircraft exhaust sulfur emissions. Geophys. Res. Lett. 23, 3603-3606.

Busen, R., and U. Schumann. 1995. Visible contrail formation from fuels with different sulfur content. Geophys. Res. Lett. 22, 1357-1360.

Busen, R., B. Karcher, A. Petzold, F. P. Schroeder, and U. Schumann. 1998. Sulfur content of aircraft fuel: influence on formation and microphysical properties of contrails. EOS, 79, n17 , S39.

Chagnon, S.A. 1981. Midwestern cloud, sunshine, and temperature trends since 1901: Possible evidence of jet contrail effects. J. Appl. Meteorol. 20, 496-508.

Chen, Y., S.M. Kreidenweis, L.M. McInnes, D.C. Rogers, P.J. DeMott. 1998. Single particle analyses of ice nucleating aerosols in the upper troposphere and lower stratosphere. Geophys. Res. Lett. 25, 1391.

Clarke, A.D. 1993. Atmospheric nuclei in the Pacific midtroposphere: Their nature, concentration, and evolution, J. Geophys. Res. 98, 20633-20647.

Clarke, A.D., J. L. Varner, F. Eisele, R. Tanner, L. Mauldin, and M. Litchy. 1998a. Particle production in the remote marine atmsophere: Cloud outflow and subsidence during ACE-1. J. Geophys. Res. 103, 16397-16409.

Clarke, A.D., F. Eisele, V.N. Kapustin, K. Moore, R. Tanner, L. Mauldin, M. Litchy, B. Lienert, M.A. Carroll, G. Albercook. 1998b. Nucleation in the equatorial free troposphere: favorable environments during PEM-Tropics, J. Geophys. Res., in press.

Cooper, P.L., and J.P.D. Abbatt. 1996. Heterogeneous interactions of OH and HO_2 radicals with surfaces characteristic of atmospheric particulate matter. J. Phys. Chem. 100, 2249-2254.

CRC (Coordinating Research Council), Handbook of Aviation Fuel Properties, 1984. CRC Report No. 530, Society for Automotive Engineers, Inc., Warrendale, PA.

Danilin, M.Y., J.M. Rodriguez, and M.R. Anderson. 1997. Aerosol particle evolution in an aircraft wake: Implications for the high-speed civil transport fleet impact on ozone. J. Geophys. Res. 102, 21453.

DeMore, W.B., S.P. Sander, D.M. Golden, R.F. Hampson, M J. Kurylo, C.J. Howard, A.R. Ravishankara, C.E. Kolb, and M.J. Molina. 1997. Chemical kinetic and photochemical data for use in stratospheric modeling. Evaluation No. 12. JPL publication 97-4, Jet Propulsion Laboratory, California Institute of Technology, Pasadena, Calif.

DeMott, P. J., D.C. Rogers, S.M. Kreidenweis, Y. Chen, C.H. Twohy, D. Baumgardner, A.J. Heymsfield, and K.R. Chan. 1998. The role of heterogeneous freezing nucleation in upper tropospheric clouds: Inferences from SUCCESS. Geophys. Res. Lett. 25, 1387.

Dentener, F.J., and P.J. Crutzen. 1993. Reaction of N_2O_5 on tropospheric aerosols: Impact on global distribution of NO_x, O_3, and OH. J. Geophys. Res. 98, 7149-7163.

Fahey, D.W., E.R. Keim, E.L. Woodbridge, R.S. Gao, K.A. Boering, B.C. Daube, S.C. Wofsy, R.P. Lohmann, E.J. Hintsa, A.E. Dessler, C.R. Webster, R.D. May, C.A. Brock, J.C. Wilson, R.C. Miake-Lye, R.C. Brown, J.M. Rodriguez, M. Lowenstein, M.H. Profitt, R.M. Stimpfle, S.W. Bowen, and K.R. Chan. 1995a. In situ observations in aircraft exhaust plumes in the lower stratosphere at mid-latitudes. J. Geophys. Res. 100, 3065-3074.

Fahey, D.W., E.R. Keim, K.A. Boering, C.A. Brock, J.C. Wilson, S. Anthony, T.F. Hanisco, P.O. Wennberg, R.C. Miake-Lye, R.J. Salawitch, N. Louisnard, E.L. Woodridge, R.S. Gao, S.G. Donnelly, R.C. Wamsley, L.A. Del Negro, B.S. Daube, S.C. Wofsy, C.R. Webster. R.D. May, K.K. Kelly, M. Loewenstein, J.R. Podolske, and K.R. Chan. 1995b. Emission measurements of the Concorde supersonic aircraft in the lower stratosphere. Science 270, 70-74.

Fried, A., B. Henry, J.G. Calvert, and M. Mozurkewich. 1994. The reaction probability of N_2O_5 with sulfuric acid aerosols at stratospheric temperatures and compositions. J. Geophys. Res. 99, 3517-3531.

Friedl, R. (ed.). 1997. Atmospheric Effects of Subsonic Aircraft: Interim Assessment Report of the Advanced Subsonic Technology Program, Reference Publication 1400, National Aeronautics and Space Administration, Washington, D.C.

Gayet, J.F., G. Febvre, G. Brogniez, H. Chepfer, W. Renger, and P. Wendling. 1996. Microphysical and optical properties of cirrus and contrails. J. Atmos. Sci. 53, 126-138.

Goodman, J., R.F. Pueschel, E.J. Jensen, S. Verma, G.V. Ferry, S.D. Howard, S.A. Kinne, and D. Baumgardner. 1998. Shape and size of contrail ice particles, Geophys. Res. Lett. 25, 1327.

Hanson, D.R., J.B. Burkholder, C.J. Howard, and A.R. Ravishankara. 1992. Measurement of OH and HO_2 radical uptake coefficients on water and sulfuric acid surfaces. J. Phys. Chem. 96, 4979-4985.

Holton, J.R., P.H. Haynes, M.E. McIntyre, A.R. Douglass, R.B. Rood, and L. Pfister. 1995. Stratosphere-troposphere exchange. Rev. Geophys. 33, 403-439.

Hu, J.H., D. Czizco, and J. Abbatt. 1997. Reaction probabilities for N_2O_5 hydrolysis on sulfuric acid and ammonium sulfate aerosols at room temperature. J. Phys. Chem. 101, 871.

Jensen, E. J., O.B. Toon, A. Tabazadeh, G.W. Sachse, B.E. Anderson, K.R. Chan, C.W. Twohy, B. Gandrud, S.M. Aulenbach, A. Heymsfield, J. Hallett, and B. Gary. 1998. Ice nucleation processes in upper tropospheric wave-clouds observed during SUCCESS, Geophys. Res. Lett. 25, 1363.

Kärcher, B. 1995. A trajectory box model for aircraft exhaust plumes. J. Geophys. Res. 100, 18835-18844.

Kärcher B. 1997. Heterogeneous chemistry in aircraft wakes: Constraints for uptake coefficients. J. Geophys. Res. 102, 19119-19135.

Kärcher , B., Th. Peter, U.M. Biermann, and U. Schumann. 1996. The initial composition of jet condensation trails. J. Atmos. Sci. 53, 3066-3083.

Kley, D., H. Geiss, and V.A. Mohnen. 1994. Concentration and trends of tropospheric ozone and precursor emissions in the USA and Europe. In The Chemistry of the Atmosphere: Its Impact on Global Change. J. G. Calvert, ed. Blackwell Scientific Publications, Oxford, U.K., pp. 245-259.

Lary, D.J., A.M. Lee, R. Toumi, M. Newchurch, M. Pirre, and J.B. Renard. 1997. Carbon aerosols and atmospheric photochemistry. J. Geophys. Res. 102, 3671-3682.

Lawson, R. P., A.J. Heymsfield, S.M. Aulenbach, and T.L. Jensen. 1998. Shapes, sizes, and light scattering properties of ice crystals in cirrus and a persistent contrail during SUCCESS. Geophys. Res. Lett. 25, 1331.

Lee, S., H. Akimoto, H. Nakane, S. Kurnosenko, and Y. Kinjo. 1998. Lower tropospheric ozone trend observed in 1989-1997 at Okinawa, Japan. Geophys. Res. Lett., 25, 1637-1640.

Liou, K.-N., S.C. Ou, and G. Koenig. 1991. An investigation on the climatic effect of contrail cirrus. In Air Traffic and the Environment—Background, Tendencies, and Potential Global Atmospheric Effects. U. Schumann, ed. Springer-Verlag, Heidelberg, pp. 154-169.

Liou, K. N., P. Yang, Y. Takano, K. Sassen, T. Charlock, and W. Arnott. 1998. On the radiative properties of contrail cirrus. Geophys. Res. Lett. 25, 1161.

Machta, L., and T. Carpenter. 1971. Trends in high cloudiness at Denver and Salt Lake City. In Man's Impact on Climate. W.H. Matthews, W.W. Kellogg, and G.D. Robinson, eds. MIT Press, Cambridge, Mass., pp. 401-405.

Mayfield, H.T. 1996. JP-8 Composition and Variability. Technical Report AL/EQ-TR-1996-0006, Armstrong Laboratory (EQL), Tyndall AFB, Fla.

Menzies, R.T., and D.M. Tratt. 1997. Airborne lidar observations of tropospheric aerosols during the Global Backscatter Experiment (GLOBE): Pacific circumnavigation missions of 1989 and 1990. J. Geophys. Res. 102, 3701-3704.

Miake-Lye, R.C., M. Martinez-Sanchez, R.C. Brown, C.E. Kolb. 1993. Plume and wake dynamics, mixing, chemistry behind a high speed civil transport aircraft, J. Aircraft. 30, 467-479.

Miake-Lye, R.C., M. Martinez-Sanchez, R.C. Brown, and C.E. Kolb. 1994. Calculations of condensation and chemistry in an aircraft contrail, in Impact of Emissions from Aircraft and Spacecraft upon the Atmosphere. Proceedings of an International Scientific Colloquium, Koln, Germany, April 18-20, 1994, U. Schumann and D. Wirzel, eds., pp. 106-112.

Minnis, P., D.F.Young, D.P. Garber, L. Nguyen, W.L. Smith, Jr., and R. Palikonda, 1998. Transformation of contrails into cirrus during SUCCESS. Geophys. Res. Lett. 25, 1157.

Mozurkewich, M., P.A. McMurry, A. Gupta, and J.G. Calvert. 1987. The mass accommodation coefficient of HO_2 radicals on aqueous aerosols. J. Geophys. Res. 92, 4163-4170.

NRC (National Research Council). 1997a. Interim Review of the Subsonic Assessment Project. Panel on Atmospheric Effects of Aviation. National Academy Press, Washington, D.C., 34 pp.

NRC. 1997b. An Interim Assessment of AEAP's Emissions Characterization and Near-Field Interactions Elements. Panel on Atmospheric Effects of Aviation. National Academy Press, Washington, D.C., 15 pp.

NRC. 1998. The Atmospheric Effects of Stratospheric Aircraft Project. An Interim Review of Science and Progress. National Academy Press, Washington, D.C., 56 pp.

Odum, J.R., T.P.W Jungkamp, R.J. Griffin, R.C. Flagan, and J.H. Seinfeld, 1997. The atmospheric aerosol-forming potential of whole gasoline vapor. Science 276, 96-99.

Penner, J., and T. Novakov. 1996. Carbonaceous particles in the atmosphere: A historical perspective to the Fifth International Conference on Carbonaceous Particles in the Atmosphere; J. Geophys. Res. 101, 19373-19378.

Ponater, M., S. Brinkop, R. Sausen, and U. Schumann. 1996. Simulating the global atmospheric response to aircraft water vapour emissions and contrails: A first approach using a GCM. Ann. Geophys. #14, 957-959.

Pueschel, R.F., K.A. Boering, S. Verma, S.D. Howard, G.V. Ferry, J. Goodman, D.A. Allen, and P. Hamill. 1997. Soot in the lower stratosphere: Pole to pole variability and contributions by aircraft. J. Geophys. Res. 102, 13113-13118.

Reichardt, J., A. Ansmann, M. Serwazi, C. Weitkamp, and W. Michaelis. 1996. Unexpectedly low ozone concentrations in midlatitude tropospheric ice clouds: A case study. Geophys. Res. Lett. 23, 1929-1932.

Rogers, D.C., P.J. DeMott, S.M. Kreidenweis, and Y. Chen. 1998. Measurements of ice nucleating aerosols during SUCCESS. Geophys. Res. Lett. 25, 1383.

Sassen, K. 1997. Contrail-cirrus and their potential for regional climate change. Bull. Am. Meteor. Soc. 78, 1885-1903.

Sassen, K., and C. Hsueh. 1998. Contrail properties derived from high-resolution polarization lidar studies during SUCCESS. Geophys. Res. Lett. 25 ,1165

Sausen, R., B. Beneberg, and M. Ponater. 1997. Climatic impact of aircraft induced ozone changes. Geophys. Res. Lett. 24, 1203-1206.

Schlager, H., P. Konopka, P. Schulte, U. Schumann, H. Zereis, F. Arnold, D. Hagen, P. Whitefiled, and J. Ovarlez. 1997. In situ observations of air traffic emission signatures in the North Atlantic flight corridor. J. Geophys. Res. 102, 10739-10750.

Schumann, U., J. Ström, R. Busen, R. Baumann, K. Gierens, M. Krautstrunk, F.P. Schröder, and J. Stingl. 1996. In situ observations of particles in jet aircraft exhausts and contrails for different sulfur-containing fuels. J. Geophys. Res. 101, 6853-6869.

Solomon, S., S. Borrmann, R.R. Garcia, R. Portmann, L. Thomason, L.R. Poole, D. Winker, and M.P. McCormick. 1997. Heterogeneous chlorine chemistry in the tropopause region. J. Geophys. Res. 102, 21411-21429.

Stephens, S.L., J.W. Birks, and J.G. Calvert. 1989. Ozone as a sink for atmospheric carbon aerosols today and following nuclear war. Aerosol Sci. Technol. 10, 326-331.

Szyrmer, W., and I. Zawadzki, 1997. Biogenic and anthropogenic sources of ice-forming nuclei: A review. Bull. Am. Meteorol. Soc. 78, 209-228.

Taleb, D., R. McGraw, P. Mirabel. 1997. Time lag effects on the binary homogeneous nucleation of aerosols in the wake of an aircraft. J. Geophys. Res. 102, 12885.

Travis, D.J., A.M. Carleton, and S.A. Chagnon. 1997. An empirical model to predict widespread occurrences of contrails. J. Appl. Meteorol. 36, 1211-1220.

Turco, R.P., and F. Yu. 1997. Aerosol invariance in coagulating, expanding aerosol plumes. Geophys. Res. Lett. 24, 1223-1226.

Twohy, C. H., and B.W. Gandrud. 1998. Electron microscope analysis of residual particles from aircraft contrails. Geophys. Res. Lett. 25, 1359

Yu, F., and R. Turco. 1997. The role of ions in the formation and evolution of particles in aircraft plumes. Geophys. Res. Lett. 24, 1927-1930.

Zhao, J., and R. P. Turco. 1995. Nucleation simulations in the wake of a jet aircraft in stratospheric flight. J. Aerosol Sci. 26, 779-795.

Acronyms

ACE	Aerosol Characterization Experiment
AEAP	Atmospheric Effects of Aviation Project
AESA	Atmospheric Effects of Stratospheric Aircraft project
BCA	Black carbon aerosol
CCN	Cloud condensation nuclei
CRC	Coordinating Research Council
FSC	Fuel sulfur content
GCM	General-circulation model
GIM	Global Interpretation and Modeling
GISS	Goddard Institute for Space Studies (NASA)
GLOBE	Global Backscatter Experiment
GMI	Global Modeling Initiative
IGAC	International Global Atmospheric Chemistry
IN	Ice nuclei
IPCC	Intergovernmental Panel on Climate Change
NASA	National Aeronautics and Space Administration
NMHC	Non-methane hydrocarbon
NRC	National Research Council

| OASTT | Office of Aeronautics and Space Transportation Technology |
| OES | Office of Earth Sciences (NASA) |

PAEAN	Panel on Atmospheric Effects of Aviation
PEM	Pacific Exploratory Mission
PSC	Polar stratospheric cloud

SASS	Subsonic Assessment Project
SONEX	SASS Ozone and Nitrogen Oxide Experiment
SUCCESS	Subsonic Aircraft: Contrail and Cloud Effects Special Study
STRAT	Stratospheric Tracers of Atmospheric Transport

| TRACE | Transport and Atmospheric Chemistry Near the Equator |